Bridging the Skills Gap in Chemical Engineering 2021

Held at the 2021 AIChE Annual Meeting

Boston, Massachusetts, USA and Online
7 - 11 November and 15 - 19 November 2021

ISBN: 978-1-7138-5683-2

Bridging the Skills Gap in Chemical Engineering 2021

Held at the 2021 AIChE Annual Meeting

Boston, Massachusetts, USA and Online
7 - 11 November and 15 - 19 November 2021

AIChE
The Global Home of Chemical Engineers

Printed from e-media with permission by:

Curran Associates, Inc.
57 Morehouse Lane
Red Hook, NY 12571

AIChE
120 Wall Street, FL 23
New York, NY 10005-4020

Phone: (800) 242-4363
Fax: (203) 775-5177

www.aiche.org

Additional copies of this publication are available from:

Curran Associates, Inc.
57 Morehouse Lane
Red Hook, NY 12571 USA
Phone: 845-758-0400
Fax: 845-758-2633
Email: curran@proceedings.com
Web: www.proceedings.com

TABLE OF CONTENTS

635087 Commercialization of Aptamer-Based Electrochemical Sensors

Abstract:

QSM Diagnostics is in the process of commercializing electrochemical sensor technology developed by the Goluch Group at Northeastern University. The initial application of the technology is for rapidly diagnosing and monitoring bacterial infections in companion animals. This is accomplished by detecting quorum sensing molecules (QSMs) that are present in specimens collected from the animals. The platform technology, however, can be adapted to almost any target molecule of interest by simply exchanging the recognition element that is attached to the sensor. The QSM sensing platform uses aptamers as recognition elements in conjunction with square wave voltammetry. Theoretically, any properly developed aptamer can be integrated to create a custom rapid, quantitative, testing system.

As an example of this turn-key solution capability, we have now demonstrated the detection of moxifloxacin, a broad-spectrum fluoroquinolone antibiotic, in human serum samples. This presentation will describe the process by which the new sensor was developed and tested as well as the challenges associated with the development process.

635137 (Invited Talk) Developing Nanosensor Technology to Detect Cancer in Patients

Abstract:

We have been developing and translating a technology to detect biomarkers quantitatively in humans to enable the early diagnosis of ovarian cancer. There is no effective screening test for ovarian cancer and no reliable method of early detection. Approximately 85% of patients with ovarian cancer will have widely disseminated peritoneal disease at presentation, resulting in five-year survival rates of < 30%. This is in stark contrast to patients diagnosed with early-stage localized disease, where survival rates are over 90%. The early detection of ovarian cancer would markedly increase survival rates. Serum biomarkers have repeatedly failed to identify early disease or change mortality rates, in part due to low specificity. We believe that the measurement of biomarkers using an implantable device in the uterine cavity will allow for discrimination of benign states from malignant disease. We are working in collaboration with clinicians to develop an implantable sensor platform for protein biomarkers capable of minimally-invasive, highly-sensitive, reproducible, in vivo measurements in disease models. We demonstrated a device to implant the sensors in animals and resected human uteri. Our goal is to develop the capability for real-time and progressive patient-centered monitoring of biomarkers for early ovarian cancer detection in at-risk patients.

630605 (Invited Talk) Wearable Health Sensors

Abstract:

The need for well-managed hydration and the monitoring of physiological parameters in real-time is not only crucial for athletes to optimize performance but to also prevent illness and injury from occurring. Understanding sweat loss and calculating hydration requirements to avoid dehydration or overhydration is a significant challenge for athletes. This is because sweat losses vary significantly from athlete to athlete both in terms of concentration of electrolytes and total volume of sweat lost. When sweating, an individual may lose as little as ~200mg of sodium per liter or as much as ~2000 mg. This variance, combined with significant differences in sweat volume, lead to vastly different hydration requirements among individuals. The RooSense Optimal Sweat Performance (OSP) Band is the first lightweight fabric sensor to provide real-time information regarding hydration levels during exercise or training through selective determination of sodium levels, a better marker for athletic performance, safety, and injury.

Monitoring sweat is a compelling choice to gain insight into a person's health at the molecular level, as sweat is rich in physiological and metabolic information that can be obtained non-invasively. In spite of advances made towards the detection of biomarkers in sweat, there is no sensor capable of long-term detection in constricted or load-bearing applications where other flexible plastic sensors might cause discomfort. Textile-based and nanocomposite technologies, such as those developed at RooSense LLC, provide an optimum scaffold for monitoring biomarkers at the surface of the skin as they do not influence the sweat composition or sweat rate and can be seamlessly integrated for wearable sensing and translational medicine applications.

This presentation will focus on the development of a prototype wearable hydration monitor, validation of the prototype in the laboratory, on body testing in an exercise physiology laboratory.

621410 Teaching Process Data Analytics and Machine Learning

Abstract:

Teaching chemical engineers to be effective in data analytics is as important today than in any time in the history of the discipline, and advances in sensor technologies, machine learning, and associated software have enabled applications that were not formerly possible. This abstract describes experiences with teaching process data analytics and machine learning over the last few years, including in (1) a new chemical engineering elective course in which the students come from chemical engineering, mechanical engineering, and engineering management, (2) a professional education course cotaught by chemical and mechanical engineering faculty whose learners primarily come with chemistry and chemical engineering backgrounds and many years of industrial experience, and (3) an undergraduate chemical engineering concentration in process data analytics with courses coming from a mixture of multiple disciplines including chemical engineering, computer science, economics, management. The overview will describe the challenges in teaching data science to chemical engineers and scientists and engineers working in the chemical and biotechnology industries, and strategies for overcoming those challenges.

622120 Teaching Big Data Science to Undergraduate Students in the University of Iowa

Abstract:

In this presentation, we will provide an overview of our course development and our experiences in teaching big data science and analytical skills to the undergraduate students in chemical engineering in the University of Iowa. These courses include: (a) Machine Learning (ML) and Artificial Intelligence (AI) course; (b) statistical analysis course; and (c) thermodynamic course. They are designed to be offered to students from freshman to junior years. We found that in general, the chemical engineering students are not getting used to use programming to analyze the data, in part because many courses they are taking focus more on the lab experiment while the analysis of the lab results is elementary and often done in Excel or by calculator. However, the students are willing to learn new analytical skills, and provided with well-designed teaching materials and problems, they can grasp the big data skills quickly. While the outcome of our experience has been encouraging, we also found there is clear lag in the field of chemical engineering to train our next generation with big data tools. This lag is part is due to the institutional tradition that emphasizes on the lab experiments and often deterministic way of thinking (with less consideration of probabilities), and in part due to the slowness in re-design the curriculum to reflect new development and ease access of ML and AI. We will provide examples to support the findings above, share the lessons we learned, and provide thoughts for next steps.

623878 Textile: Tutorials in Experimentalist Interactive Learning

Abstract:

Background:

Confocal microscopy is a common preclinical imaging modality to study a range of tissue features. Fluorescent image microscopy can provide insights into cell density, morphology, and cell-cell relationships. In our lab, we use immunofluorescent staining to investigate changes in cell and microenvironment features in response to development, injury, and treatment in the brain. We have developed image processing pipelines that threshold, segment, and quantify cell features. More recently, we have incorporated data science-enabled analysis and techniques into our imaging workflow. While the lab is comprised of researchers who are wet-lab experimentalists and data-driven analysts, only the second group has formal data science training. Wet-lab experimentalists can significantly benefit from having data science literacy to apply data science methodologies to their imaging datasets. However, there is no formal training regime to train wet-lab experimentalists in lab-specific data science methodologies. To address this training gap, we developed Tutorials in EXperimentalisT Interactive LEarning (TEXTILE) for data-science enabled fluorescent image processing.

TEXTILE is a module-based, semi-linear educational system developed with laboratory-based learning in mind. Modules combine data science fundamentals with specific biological imaging processing techniques. For example, we pair a data science module about data management with an image processing module on quantitative cell morphology characterization. The modular format allows us to train students on managing experimental data and pair that with quantifying features of cells that are relevant to their specific project of interest. Additionally, we designed TEXTILE such that the learning process is adaptable to starting skill level. Learning paths are based on entry knowledge but allow students to pick and choose their modules based on interest or previous experience. For example, someone with ample coding knowledge can skip the introductory Python modules and begin with the experimental image processing modules. Finally, we designed TEXTILE for a variety of degree levels. As a laboratory at a large public university, we engage with undergraduate and high school students, in addition to master's thesis and doctoral thesis students. We accomplished degree-level accessibility through the semi-linear system that assumes no prior knowledge in our research area. Since we already use TEXTILE to train high school students participating in laboratory research that have no previous knowledge of the subject topic, we aim to include more local high school students and expand TEXTILE for science education and outreach as well as in-laboratory education.

Methods:

TEXTILE is composed of topic-specific modules. Each module is designed around a data science topic, lasts an hour and a half, and contains four sections: the pre-module activity, main module, post-module reflection, and feedback form (Figure 1a). The instructor sends the pre-module activity to students for self-driven learning a week before teaching the main module. The pre-module's purpose is to prime students for the module's main lesson with a high-level activity. For example, in our data management module, students analyze a grocery shopping trip and decide where to store specific categories of food and perform a space evaluation on how much room in each storage location the food will take up.

The main module aims to teach the bulk material in digestible lectures broken up by interactive sessions. The main module's detailed structure consists of six parts: introduction presentation, interactive session one, main presentation workshop, interactive session two, concluding presentation, and a reflection session (Figure 1b). The introduction presentation introduces the module's topic, outlines the structure the main module will follow, and leads into the first interactive session. Interactive session one guides the student through an exploration of the module material. For coding-based modules, instructors lead interactive sessions with Jupyter Notebook. During interactive session one, the instructor walks students through the code's basics as students code along. After the students have written the code, the instructor begins the main presentation, which has two purposes: (1) to increase the students' connection between the pre-module activity and the module introduction and (2) to dive deeper into the technical aspects of the first interactive session. Once students understand the technical elements of the main module, they begin interactive session two, where they apply technical details of the lesson to further explore the code written in the earlier interactive session. After students have an opportunity to explore, the instructor gives the concluding presentation, which includes wrapping up the main takeaways, emphasizing the module's motivation, and thanking students for their participation.

Finally, the main module ends with a prompt for students to reflect on what they have learned and how their opinions about the module's topic have changed since the pre-module activity. The post-module reflection is completed individually by each student a day or two after the main module. The purpose is to provide a student-led review activity that encourages them to explore content from the main module on their own time to solidify concepts and build confidence. After the whole module is completed, students are sent a feedback form through Google Forms. With the feedback form, students reflect on their growth and the impact of each component of the module. The feedback is then implemented back into the module for the next cohort of students.

Results:

In the summer of 2020, the first cohort of students went through six TEXTILE modules of paired data science and image processing techniques used within the lab. The first cohort had 15 students: six high school students, four undergraduates, and five graduate students. The first cohort was taught live by a Ph.D. student from the lab's data-driven analysis team. The topics covered included GitHub and version control, introduction to data management, introduction to experimental data management specific to the lab, intro to image processing, thresholding, image segmentation, cellular morphology, and quantifying cell shape features. By the end of the first cohort, all students completed every TEXTILE module and the high school and undergraduate students developed individual projects using protocols learned over the summer. They each created Jupyter Notebooks for a self-led project with a brain cell morphology lit review, data management plan, and image processing pseudo-code. Many students from the summer 2020 cohort chose to continue researching with the lab and are currently developing these projects further.

In fall 2020, a small cohort of two undergraduate students explored the TEXTILE modules asynchronously. The students watched a recording of the main module and participated in asynchronous discussions with previous students and a Ph.D. student-instructor via Slack. We also provided these students with published image data that no one previously analyzed for cell morphology features. The students are currently completing retroactive analysis on this data using techniques learned from various TEXTILE modules.

For summer 2021, TEXTILE is expanded to include three modules on machine learning and two additional modules on experimental image acquisition and manual cell counting, incorporating student feedback from 2020. The main takeaways from student feedback included developing a formalized mentoring process for student projects and teaching via a hybrid live and asynchronous format.

Conclusions:

TEXTILE is a successful data science educational system for teaching data science techniques commonly used in experimental research labs. The TEXTILE system works for protocol training of fellow lab members, onboarding new students, and science education outreach. The combination of interactive sessions with instructor-led presentations gives students ample opportunity to work hands-on with real data from the lab in a guided environment. Students from the first cohort had the opportunity to develop their cell morphology research projects. In contrast, students from the second cohort took the lessons learned from TEXTILE and applied them for retroactive analysis of previously published data. The TEXTILE system is built to be robust and adaptable so that other laboratories can follow the methodology to develop training modules for their own data science techniques.

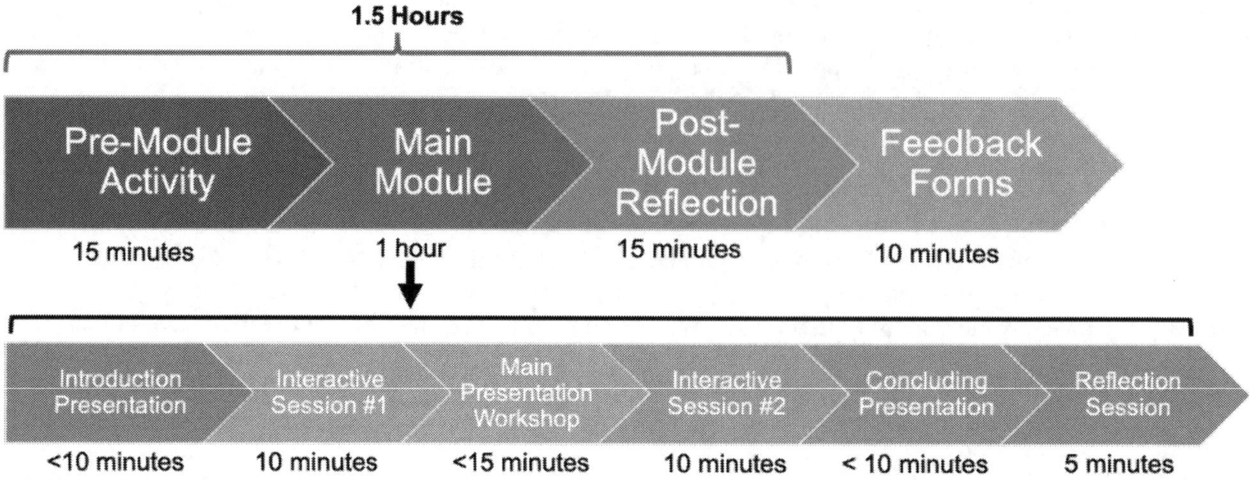

Figure 1. (a) Top row: TEXTILE process composed of four parts – pre-module activity, main modules, post-module reflection, and feedback forms. (b) Bottom Row: The six sections of the main module - introduction presentation, interactive session #1, main presentation workshop, interactive session #2, concluding presentation, reflection session.

622481 Teaching Artificial Intelligence to Chemical Engineers: Experiences from a 35-Year-Old Course

Abstract:

Given the recent success of artificial intelligence (AI) in a wide variety of applications, it has become quite clear that chemical engineering students need to become familiar with applied AI techniques [1]. There is great interest in the academic community in developing a course to address this important need. I have been teaching such a course to chemical engineering undergraduate and graduate students since 1986 [2], first at Columbia for three years, then at Purdue for about twenty years, and back again at Columbia now. It was mostly on *symbolic AI* in the early years, with an emphasis on *knowledge-based expert system* concepts and techniques, for various applications in chemical engineering such as design, diagnosis, control, optimization, and process safety. From the 1990s onwards, as machine learning techniques started showing promise, my course evolved to include topics such as clustering, neural networks, statistical classifiers, graph-based models, and genetic algorithms. The current version at Columbia has the first one-third of the course on *symbolic AI* and the rest on machine learning-based *numeric AI* concepts and techniques. It is mainly a project-based course where the students develop systems using AI techniques for chemical engineering applications. In the early years, the students programmed in AI languages such as LISP and OPS5, but now using the easier tools such as Python and TensorFlow. In this presentation, I will discuss my experiences in teaching this course, including the syllabus and the projects we have done over the decades.

Reference:

1. Venkatasubramanian, "The Promise of Artificial Intelligence in Chemical Engineering: Is it here, finally?", *Perspective article, AIChE Journal,* 65(2), pp. 466-478, 2019.

- Venkatasubramanian, V. "Artificial Intelligence in Process Engineering: Experiences from a Graduate Course," *Chemical Engineering Education*, Fall Graduate Education Issue, November 1986, pp. 188 – 192.

623762 Insights from Teaching "Data Analytics for Chemical Engineers"

Abstract:

In this talk I will discuss the origins and current status of the "Data Analytics for Chemical Engineers" course developed and taught at Georgia Tech. This course is designed to provide chemical engineers with a broad introduction to data science and machine learning techniques including regression, classification, clustering, and dimensional reduction. The course is currently one of two core courses in the "Data Science for the Chemical Industry" (DSCI) certificate program, a unique program that combines undergraduate, graduate, and professional education. The advantages and challenges of teaching data science to chemical engineering students at varying stages of their careers will be presented. Some pedagogical and technical aspects of the course, including the semester-long group project, the use of the Python programming language, and teaching in a fully online environment will be discussed.

621458 Tips and Pitfalls to Avoid When Teaching Machine Learning with Python to Chemical Engineering Students at the Undergraduate and Graduate Level

Abstract:

Every area of Chemical Engineering research and industry, including Catalysis, Biomolecular Engineering, Cellular Engineering, Computing and Simulation, Nanotechnology, Materials, Polymers, Sustainable Energy, and Microfabricated Systems, stands to be transformed by data science—the ability to produce, mine, build models, and learn from vast quantities of data. However, many chemical engineers have little to no training in data science and machine learning. In the Department of Chemical Engineering at University of Michigan, I have recently created and taught two classes on Data Science and Machine Learning to graduate and undergraduate chemical engineering students. The classroom material is aimed to prepare students to use modern data science tools (e.g., Python programming and the Numpy, scikit-learn, matplotlib, and pandas libraries) to analyze data that engineers may encounter. Course content expands the students' horizons of the utility of data science in real-world context and enhances written and oral communication abilities as well as teamwork. Topics and applications covered include data collection, curation, and supervised and unsupervised machine learning. Algorithms covered will include the perceptron, adaline, logistic regression, gradient descent, principal component analysis, random forests and feed forward neural networks. Homework and lab exercises include hands-on practice of using data science and machine learning with Python in Google Colab notebooks. Students are responsible for a team-based data science / machine learning project. In this talk, I will share my experience and provide some tips and pitfalls to avoid when teaching data science / machine learning content to chemical engineering students.

633997 Problem-Based Learning on Incorporation of Data Analysis Skills into a Senior Course

Abstract:

Through years of production, chemical process industry (CPI) accumulated rich data asset. In the age of information and AI, processing facilities are using data in new ways to improve efficiency, reliability, and safety. However, skilled workforce is in shortage. To embrace the digital transformation, chemical engineering students need to get in-depth training on data analytics skills.

Monte Carlo methods defined broadly a statistical approach to provide approximate solutions to mathematically complex optimization or simulation problems facing uncertainty by using random sequences of numbers. Monte Carlo simulations are used to model the probability of different outcomes in a process that cannot easily be predicted due to the intervention of random variables. It is a technique used to understand the impact of risk and uncertainty in prediction and forecasting models. Monte Carlo methods have been widely used in chemical industry. In the application of asset health, heat exchanges' performance was modeled considering manufacturing tolerances and uncertain flow distribution.

In a "Advanced Analysis" class, the fundamentals of statistics, outlier detection, and Monte Carlo Simulation were explained. Then a real-world problem was assigned per the request from the industry. The students were divided into groups of 4-5 people each, and they were asked to figure out which heat exchangers in a refinery need to be cleaned based on condition rather than time, with the help of Monte Carlo Simulation and statistics analysis [2]. This approach was well-received by the students, some of them eventually became process engineers or maintenance engineers, who are using this skill in their work.

623042 Chemical Engineering Analysis through Systematic Optimization

Abstract:

With the rapid development of computing technology, computer-aided design skills are in high demand. The course, CHE5031 Chemical Engineering Analysis, has been offered at the University of Minnesota Duluth since 2015. This course aims to help students build a fundamental understanding of chemical engineering systems by developing mathematical and statistical models and simulations using digital computers. It also aims to develop a systematic understanding and a critical awareness of process optimization and analysis of results.

In 2021, MATLAB was selected as the computational language in the course. As a multi-paradigm numerical computing environment and proprietary programming language developed by MathWorks, MATLAB is a powerful tool for computer-aided design and Chemical Engineering analysis.

As shown in Figure 1, the course is organized into three major sections with 10 chapters:

1. Chapters 1 and 2 introduce modeling in chemical engineering processes and MATLAB skill preparation.

2. Chapters 3 through 7 introduce some advanced algorithms and computational methods with case studies in Chemical Engineering.

3. Chapter 8 through 10 integrate algorithms and methods for tackling complicated case studies in Chemical Engineering. There are many Chemical Engineering-related examples in each chapter at both the undergraduate and graduate levels.

This course's most distinguishing feature is to apply systematic optimization for chemical engineering analysis to help students build their confidence for optimized data-driven decision-making. Students do not just stop learning on the existing optimization toolbox and global optimization toolbox in MATLAB - they gain added knowledge of the limitations of MATLAB's optimization solvers. Additional topics include enhanced optimization methods, such as Lagrange's method of converting constrained optimization to non-constrained optimization, conversion of inequality constraints to equality constraints, the penalty function method, and an enhanced constraints method to avoid the unwanted "zero" local optimal point.

The optimization techniques are demonstrated by case studies in model fitting, solving complex equations, and process optimization. Students also learn optimal experimental design to research influencing factors, their levels, the significance of each factor, and their combined effects. Besides the in-class exercises, students have four main assignments:

1) Model fitting of a particle size distribution via different ways by MATLAB apps and many optimization methods.

2) Optimization of the placement of a petrol station located on the road among 5 towns and further questions with weighted optimization, and with both equality and inequality constraints.

3) Optimization of a nodal network in a Cartesian grid to simulate steady-state 2D conduction heat transfer without internal heat resources.

4) Application of orthogonal experimental design for finding optimal parameters that minimize the operation costs in solvent extraction flowsheets using a ternary diagram.

Our Chemical Engineering Analysis course bridges the skills gap for our graduate students and some undergraduate students who selected this course, especially enabling them to use the "data science" techniques, skills, and modern engineering tools necessary for engineering practice.

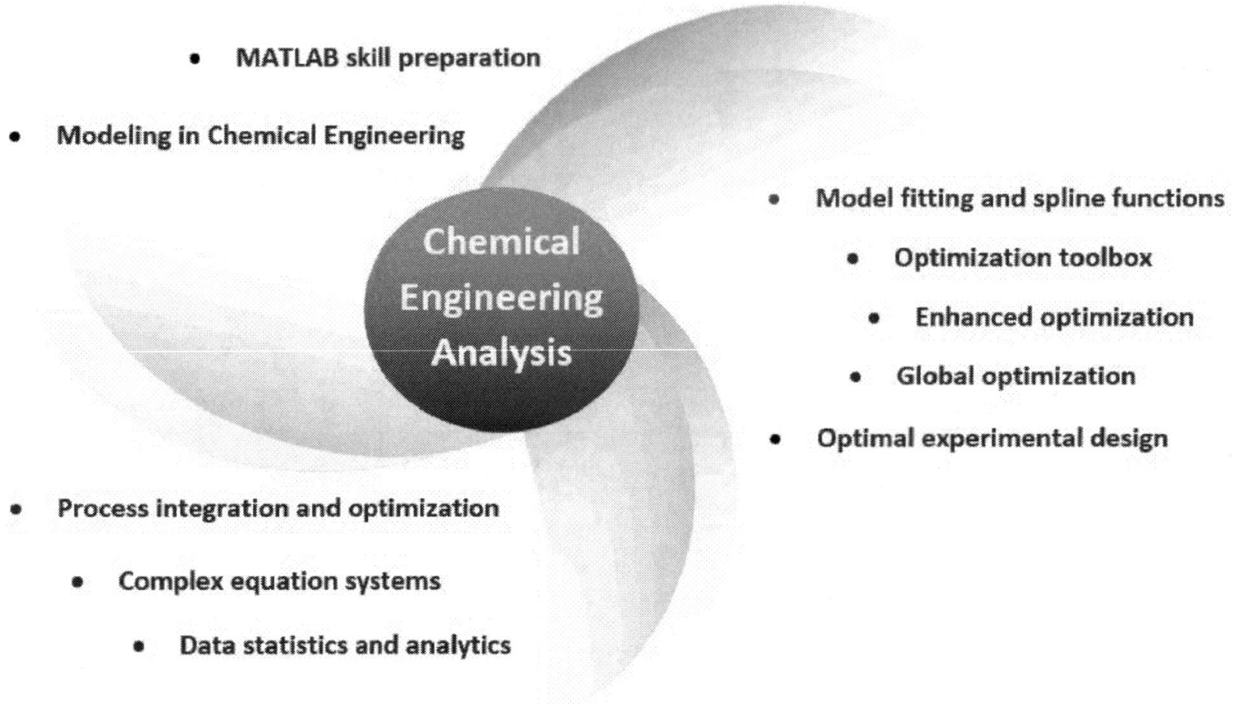

Figure 1: Three major sections with 10 chapters in the Chemical Engineering Analysis course

632884 Teaching Domain Experts Data Science: A Progress Report from a Purdue Initiative

Abstract:

Domain knowledge plays an underemphasized but nonetheless critical role in nearly all stages of successful applications of data science and machine learning. From data collection to application implementation—with curation, problem featurization, model design, and validation, among other ingredients, in between—domain knowledge informs countless decisions made by practitioners regarding what to collect, how to collect it, how to model it, and how to evaluate performance. In the context of chemical engineering education, this means that training opportunities need to exist that enable domain experts to learn the language of modern data science, including its limitations and pitfalls. In response to this, two years ago we developed a data science course for chemical engineers with the goal of training data fluent chemical engineers. In this presentation, I'll cover what we have done, what has worked, and what has failed based on the last three offerings of our course. Among the salient observations is that a sizeable group of chemical engineering undergraduates are already eager for course offerings along these lines and are self-educating as they are able. We also observe that the domain experts in the course are quick to realize how certain modeling assumptions can lead to unphysical consequences. These are very hopeful signs. We also observe a gap between programming experience and data science needs for a large fraction of students that are otherwise enthusiastic about this topic. This is a more difficult gap to bridge in a single course, and motivates some suggestions at the end of the presentation on what training is required for chemical engineers to effectively work on data science problems.

629782 Deployment of Machine Learning Models in Pharmaceutical Development

Abstract:

The advantage of model driven decisions in research organizations has long been recognized and applied in the pharmaceutical research and development. From the conceptualization of the Quality by Design framework as a series of mathematical constructs [1,2] to the application of digital twins [3]. Machine Learning methodologies are widely applicable and efficient tools to construct mathematical models of relevant pharmaceutical processes enabling process optimization, robustness characterization, and failure rate estimation. However, the widespread adoption of these approaches requires a combination of streamlined availability, and institutional know-how to interpret and incorporate the results in the decision-making process.

This talk addresses different approaches to leverage emerging and stablished tools in Machine Learning to pharmaceutical development applications by comparing the adoption of specific algorithms and computational platforms. Specifically, a generic workflow for data exploration analysis, design of experiments, feature selection, and model exploration has been developed and deployed with a combination of cloud computing and Jupyter notebooks. In our experience this approach has been more successful at integrating a model-based decision making culture in process development than previously attempted alternatives.

[1] Peterson, J.J., 2008. A Bayesian approach to the ICH Q8 definition of design space. *Journal of biopharmaceutical statistics*, *18*(5), pp.959-975.

[2] Garcia-Munoz, S., Luciani, C.V., Vaidyaraman, S. and Seibert, K.D., 2015. Definition of design spaces using mechanistic models and geometric projections of probability maps. *Organic Process Research & Development*, 19(8), pp.1012-1023

[3] Chen, Y., Yang, O., Sampat, C., Bhalode, P., Ramachandran, R. and Ierapetritou, M., 2020. Digital Twins in Pharmaceutical and Biopharmaceutical Manufacturing: A Literature Review. *Processes*, *8*(9), p.1088.

626098 Unsupervised Learning from Sets of Data Using Contrastive Latent Variable Models

Abstract:

In unsupervised learning, dimensionality reduction is an important tool for data exploration and visualization. Because these aims are typically open-ended, it can be useful to frame the problem as looking for patterns that are enriched in one dataset relative to another. These pairs of datasets occur commonly, for instance a population of interest vs. control or signal vs. signal free recordings.However, there are few methods that work on sets of data as opposed to data points or sequences. Here, we present a probabilistic model for dimensionality reduction to discover signal that is enriched in the target dataset relative to the background dataset. The data in these sets do not need to be paired or grouped beyond set membership. By using a probabilistic model where some structure is shared amongst the two datasets and some is unique to the target dataset, we are able to recover interesting structure in the latent space of the target dataset. The method also has the advantages of a probabilistic model, namely that it allows for the incorporation of prior information, handles missing data, and can be generalized to different distributional assumptions. We describe several possible variations of the model and demonstrate the application of the technique to de-noising, feature selection, and subgroup discovery settings.

627704 Energy Dispersive X-Ray Hyperspectral Image Analysis and Chemometrics for Catalyst Characterization

Abstract:

Multi-dimensional images with chemical information embedded at pixel-level (so-called Hyperspectral Imaging) is the next generation of Image Analysis and Machine Vision. This advanced technology has attracted the attention of different industry sectors in the last two decades, ranging from metalurgic, mining and pharma, to high-value-added manufacturing industries such as biotechnology and biomedical imaging. Recently, the combination of Scanning Transmission Electron Microscopy (STEM) and Energy Dispersive X-ray spectroscopy (EDX) has been developed to characterize the surface area of catalysts in the chemical industry. The measurements are done by scanning the electron beam over the sample and at the same time measuring the X-ray spectrum – so-called spectral imaging [1]. The spectral dimension of the hyperspectral images consists of X-ray counts acquired at different energy channels. Chemical data is corrupted with noise, which comes from the time-dependent arrival of discrete particles on the sensor. This noise typically follows a Poisson distribution, which represents the probability of occurrence or events (X-ray counts) during a given period of time.

The application of Multivariate Image Analysis (MIA) techniques and Multivariate Curve Resolution (MCR) models becomes essential for the analysis of EDX hyperspectral images. This contribution proposes a modeling framework that permits segregating hyperspectral X-Ray images into simpler images (so-called Distribution Maps, DM's), which can be directly related to each of the chemical compounds present in the mixture [2]. From these DM's, chemical-textural score images (SI's) are further obtained. Finally, from all these DM's and SI's, different types of features, such as quantitative, morphological or textural can be extracted and combined into new data structures [3]. This new source of information is used to build multivariate statistical models for process understanding and prediction purposes at a MIA image-based level. This approach allows us to study similarities and differences between and within types of catalysts -i.e. different samples of the same catalyst across batches, and across locations, and the potential effects on quality properties of interest. To illustrate the chemometric framework for catalyst characterization, STEM-EDX images of real industrial catalyst will be used.

[1] R. Haswell, D. W. McComb and W. Smith, Preparation of site-specific cross-sections of heterogeneous catalysts prepared by focused ion beam milling, Journal of Microscopy (2003) Vol. 211, Pt 2, pp. 161-166
[2] J.M. Prats-Montalbán, A. de Juan, A. Ferrer, Multivariate image analysis: a review with application, Chemometrics and Intelligent Laboratory Systems, 107: 1-23, 2011.
[3] C Duchesne, JJ Liu, JF MacGregor, Multivariate image analysis in the process industries: A review, Chemometrics and Intelligent Laboratory Systems 117, 116-128, 2012.

634389 Applications of machine learning in production enhancement and field monitoring in Oil & Gas industry

Abstract:

Due to the fast advancement of machine learning technologies and availability of diverse types of data collection platforms, we have been seeing the growing machine learning applications in a wide spectrum of industries, spanning from emerging industries, e.g., autonomous driving, to traditional ones, such as Oil & Gas (O&G).

In this talk, we will discuss three different machine learning applications in the O&G industry. In specific, we will discuss a machine learning-based approach for understanding the multiphase flow system studied in the experimental fluid mechanics; which shows the potential of developing a complimentary methodology to assist the first principal physics; secondly, we will describe a deep learning-based object detection system that can be used to monitor the operation field or the natural environment, with the goal to improve operation safety and meet the regulation requirement ; Thirdly, we will introduce the content understanding of Process & Instrument Diagrams (P&IDs) using computer vision techniques.

Each of the above studies represents a unique usage scenario in the value chain of O&G industry. We would like to demonstrate the diverse types of machine learning applications and discuss their respective technical challenges that O&G industry presents.

637411 The Hard Skills of Emotional Intelligence for Effective Engineering Leaders

Abstract:

High performing teams and leaders with emotional intelligence are more likely to achieve their goals. This workshop featuring Dr David Caruso of Yale's Center for Emotional Intelligence, focuses on the ability model of emotional intelligence and begins with a hands-on illustration of the four "hard" skills of EI, followed by demonstrations of the notion that "emotions are data" and the cognitive costs of emotion suppression. After a review of a brief case study, we then provide participants with a means to apply these skills to situations faced by leaders in engineering using the Emotional Intelligence Blueprint.

AUTHOR INDEX

AIChE
120 Wall Street, FL 23
New York, NY 10005-4020

CURRAN ASSOCIATES INC.
proceedings.com

ISBN 978-1-7138-5683-2
9 781713 856832